Theory
INTO
Practice

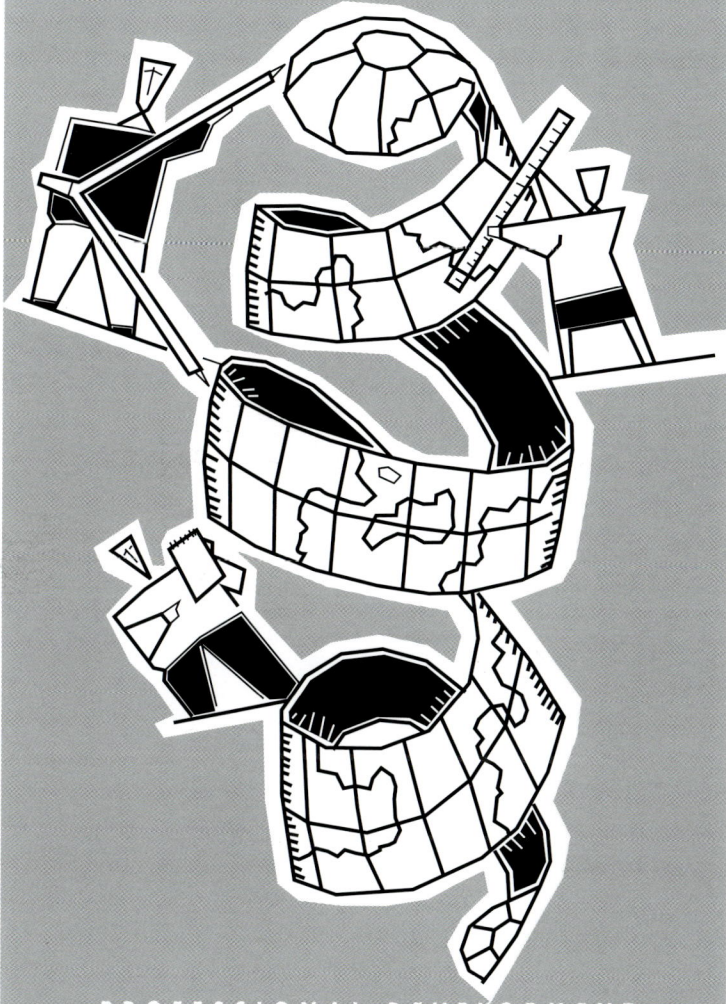

PROFESSIONAL DEVELOPMENT
FOR GEOGRAPHY TEACHERS
Series editors: Mary Biddulph and Graham Butt

Extending
Writing
Skills

GRAHAM BUTT

Geographical
Association

The author

Dr Graham Butt is Senior Lecturer in Geographical Education in the School of Education, University of Birmingham.

The series editors

Dr Mary Biddulph is Lecturer in Geography Education in the School of Education, University of Nottingham and Dr Graham Butt is Senior Lecturer in Geographical Education in the School of Education, University of Birmingham.

ISBN 1 899085 09 2
First published 2001
Impression number 10 9 8 7 6 5 4 3 2 1
Year 2004 2003 2002

Published by the Geographical Association, 160 Solly Street, Sheffield S1 4BF.
E-mail: ga@geography.org.uk
Website: www.geography.org.uk
The Geographical Association is a registered charity: no 313129.

The Publications Officer of the GA would be happy to hear from other potential authors who have ideas for geography books. You may contact the Officer via the GA at the address above. The views expressed in this publication are those of the author and do not necessarily represent those of the Geographical Association.

Designed by Ledgard Jepson Limited
Printed in Hong Kong through Colorcraft Ltd.

University of
Hertfordshire

College Lane, Hatfield, Herts. AL10 9AB

Learning and Information Services
de Havilland Campus Learning Resources Centre, Hatfield

For renewal of Standard and One Week Loans,
please visit the web site **http://www.voyager.herts.ac.uk**

This item must be returned or the loan renewed by the due date.
The University reserves the right to recall items from loan at any time.
A fine will be charged for the late return of items.

Geographical
Association

Photo: News Team International Ltd.

Contents

Photo: News Team International Ltd.

Editors' preface

Theory into Practice is dedicated to improving both teaching and learning in geography. The over-riding element in the series is direct communication with the classroom practitioner about current research in geographical education and how this relates to classroom practice. Geography teachers from across the professional spectrum will be able to access research findings on particular issues which they can then relate to their own particular context.

How to use this series

This series also has a number of other concerns. First, we seek to achieve the further professional development of geography teachers and their departments. Second, each book is intended to support teachers' thinking about key aspects of teaching and learning in geography and encourage them to reconsider these in the light of research findings. Third, we hope to reinvigorate the debate about how to teach geography and to give teachers the support and encouragement to revisit essential questions, such as:

- Why am I teaching this topic?
- Why am I teaching it in this way?
- Is there a more enjoyable/challenging/interesting/successful way to teach this?
- What are the students learning?
- How are they learning?
- Why are they learning?

This list is by no means exhaustive and there are many other key questions which geography teachers can and should ask. However, the ideas discussed and issues raised in this series provide a framework for thinking about practice. Fourth, each book should offer teachers of geography a vehicle within which they can improve the quality of teaching and learning in their subject; and an opportunity to arm themselves with the new understandings about geography and geographical education. With this information teachers can challenge current assumptions about the nature of the subject in schools. The intended outcome is to support geography teachers in becoming part of the teaching and learning debate. Finally, the series aims to make classroom practitioners feel better informed about their own practice through consideration of, and reflection upon, the research into what they do best - teach geography.

Mary Biddulph and Graham Butt
January 2001

Introduction

Establishing effective communication in the classroom is fundamental to teaching and learning geography, or indeed any subject. Some forms of communication are particular to geography - for example, maps, aerial photographs and satellite images - and these have their own collections of research. This book, however, is primarily concerned with written communication and the ways in which writing activities in the geography classroom can assist or, if used injudiciously, hinder the learning process. The main focus of this book is, therefore, the expression of geographical concepts and ideas through writing, although some attention is also given to talking, listening and reading within the geography classroom as a precursor to successful writing; indeed discussion among students is often an essential pre-requisite to the completion of good quality extended writing in geography.

A number of small-scale classroom-based research projects conducted over the past ten years in a variety of comprehensive schools in the West Midlands have provided the foundations for this book. However, the particular activities described here took place over a fourteen-week period. The overall aim of the action research has been to address a fundamental, but extremely broad question: 'How can the different uses of language in the geography classroom help young people write more effectively?'

A considerable literature already exists concerning the relationship between language, thought development and learning, some of which has been interpreted specifically for teachers of geography (see, for example, Williams, 1981; Slater, 1989; Carter, 1991; Butt, 1997). There are also several general texts that seek to help teachers promote the broader use of language in their classrooms (Andrews, 1989; Sheeran and Barnes, 1991; Wray and Lewis 1994, 1997; McCarthy and Carter, 1994; SCAA, 1997a,b). This literature has been supplemented by a variety of QCA booklets (some of which also report on research findings) which seek to promote the improvement of young people's literacy in line with recent government initiatives (QCA, 1999a,b). However, a number of significant questions concerning how students use language to learn still remain to be answered. As Lambert and Balderstone state:

> 'many questions relating to the role of language in cognitive development remain unanswered. In particular, more research is needed into the processes by which children use language both to learn and to develop their understanding of concepts through talking and writing. The difficulty in collecting and interpreting such evidence is one reason for the lack of evidence. There are also problems isolating specific learning associated with language development from other learning processes in the classroom' (2000, p. 214).

The point of this book, and indeed of all the titles in this series, is to bring theory and practice together so that teachers understand how educational research can help them to become more effective practitioners in the classroom. This book therefore has its origins in the notion that teaching is a research-based profession. It is written from the stance that the use of language is essential for communicating what is known and understood, and is also fundamental to the whole *process* of learning. It should help geography teachers become more fully aware of their role in creating the conditions in which students can successfully 'talk and write to learn'.

Photo: Liz Taylor

1: Communication in the geography classroom

It is a truism that teacher talk tends to dominate the communication that occurs within the classroom. Geography teachers are like all other subject teachers in this respect:

- we create fixed sets of rules for the ways in which we want students to communicate with us ('hands up', 'don't shout out'),

- we establish periods in each lesson when students are expected to restrict the communication they engage in ('when I'm talking you must listen to me', 'too much noise - you should all be writing now'), and

- we encourage students to communicate orally in different ways (for example, when answering 'whole class' questions, reporting back from group activities as individuals, giving mini presentations).

The times when students can 'legitimately' communicate orally are usually quite restricted and they soon learn the accepted parameters for communication in the geography classroom. When we as geography teachers consider just how important talking is to the whole learning process, we begin to realise that the opportunities we offer students to engage in subject related discussion or questioning are educationally extremely significant. If we totally dominate in the classroom and the process of oral communication becomes 'one way', then the learning experience for many students is restricted. If they are to understand fully the subject's concepts, ideas and terminology students must regularly engage in talking about geography, as well as listening to the teacher and their peers. The significant point is that talking is often the precursor both to learning and to the production of good quality extended writing (Williams, 1981; Slater, 1989). Without some opportunity to discuss the geographical tasks and concepts in front of them many students do not produce substantial pieces of sound geographical writing.

Much has been written about the ways in which teachers communicate in geography classrooms, particularly through oral question-and-answer sessions (see, for example, Roberts, 1986; Carter, 1991; Butt, 1997). Attention has also been given to the manner in which such communication may either stretch student thinking into new areas, or merely repeat a game of rote memorisation of facts, which students must regurgitate from one geography lesson to the next. The majority of questions we ask in class are 'closed' in that only one 'right' answer is possible; often we phrase such questions in such a way as

to offer our students very limited opportunities to explore new thoughts and ideas in geography. Similarly many 'recall' questions simply test what the students already know, rather than encouraging new understandings.

Why do we pose so many closed questions (both orally and in the writing tasks we expect students to complete)? I would argue that there are several reasons for this, some of which relate closely to issues of class management and others to our preparedness to take risks. Closed questions are certainly easier to frame than open questions; they can be assessed quickly and are more convenient for keeping academic 'control' of a lesson. A series of one word answers may also have the advantage of driving the pace of a lesson, involving the participation of a large number of students in the class and not threatening the teacher with the possibility of deviating into areas that he or she has not planned for. This is true for all closed oral (or written) questions, whether they are set against a 'stimulus' – a piece of text, a statistical diagram, a map – or any other resource. By comparison, 'open-ended' questions invite rather more tentative and exploratory answers, which in themselves provide evidence of fresh thinking and new learning. This may, in turn, create management and assessment problems within the geography classroom. Nonetheless, such questions are truly 'educational' in that they push students into higher order thinking and reasoning, often by making them engage in analysis, in synthesis, in decision-making and in formulating conclusions.

Marsden (1995) describes what he considers to be the features of good questioning within the geography classroom. The list below is based on his work on oral questions and answers, but some points apply equally well to written questions.

Marsden (1995) reminds teachers to:
1 ask questions fluently and precisely,
2 gear questions to the students' state of readiness,
3 involve a wide range of students in the question and answer process,
4 focus questions on a wide range of intellectual skills, and not just on recall,
5 ask probing questions,
6 not accept each answer as having equal validity,
7 sensitively redirect questions to allow accurate and relevant answers to emerge, and
8 use open-ended as well as closed questions in order to invite creative thought and value judgements.

Asking a high proportion of closed questions, either in oral or written work, may be organisationally convenient in that it enables us to get students to 'do something' without too much fuss – but it is often the case that not much real learning is going on!

The language of students

If we accept that it is important to give students opportunities to discuss the geography they are learning and to answer thought provoking open-ended questions, what does this imply about the ways in which we think geography should be taught? First, we should encourage students to use language in *exploratory* ways, rather than solely in *transactional* ways. That is, we should allow students to use personal and expressive forms of language (exploratory) which help to reveal what they think, feel and believe.

Photo: News Team International Ltd.

By contrast, transactional language, which is more formal and structured, is used to convey factual information and concepts in a logical and ordered sequence. Unfortunately teachers often expect students to be able to produce good quality transactional writing too quickly and with little support. Students must be given opportunities to 'play' with language to discover new meanings, rather than simply use it to convey final answers. As Lambert and Balderstone suggest:

'Students should be given opportunities to talk in a range of contexts and for a variety of purposes in geography including describing and explaining, negotiating and persuading, exploring and hypothesising, challenging and arguing' (2000, p. 215).

A strand of research, which has been developed around these ideas, looks at both developing thinking skills and supporting students in analysing how they learn. For example, Leat (1998, 2000), Leat and Nichols (1999) and Leat and Kinninment (2000) have explored the ways in which students 'think about their thinking' (metacognition).

These researchers have sought to analyse the language that students use when solving problems and have encouraged students to think about how they learn, so that they can approach subsequent geographical tasks and questions more effectively. Although this work has not yet been extended fully into researching students' writing, it has made some important links between thinking, talking and writing.

Second, we can offer students the opportunity to talk by using role-play and simulation activities and decision-making exercises in geography, most of which will involve group work at some stage. Some geography teachers are concerned that such activities may transfer control of the learning process from them to the students. While this is a justifiable concern, teachers must accept that it is also the only way in which students will start to learn for themselves, rather than relying directly on the teacher for instruction and guidance. It is also important to remember that:

> 'over-enthusiastic interventions [by the teacher] often takes the initiative away from the students, who should be developing an understanding of their roles and responsibilities in maintaining discussions and completing set tasks'
> (Butt, 1997, p. 159).

Ensuring that students engage constructively in more independent learning is not always easy and takes time; it is only after the fears of messy and non-directed learning are conquered that students can move towards more independent and valuable forms of learning with language. This is not to say that 'anything goes' - even though students will be offered greater space and time to discuss or write freely, properly conducted discussion work or exploratory writing still needs very clearly defined and expressed parameters.

From talking to writing

Although teachers tend to dominate oral communication in the geography classroom and often restrict opportunities for meaningful student talk, the main focus for this book is on students' writing. Two key questions present themselves in relation to this focus:

1. How can we ensure that geography students engage in high quality extended writing?
2. What is the link between writing and learning in the geography classroom?

In relation to question 1, let us start by considering what is meant by the phrase 'high quality extended writing'. Many students produce writing that is poorly structured, overly concise and unbalanced, and which is also incapable of conveying complex messages, ideas and thoughts. Although the reasons for this are numerous, often the key to the regular production of unsatisfactory extended writing lies in the nature of the task originally set by the teacher. For example, if you ask a student to name the capital of Spain or to state the term given to the movement of material by a river, you will get a one-word answer! Extended writing can, therefore, only be produced if we ask or encourage students to ask the right kinds of questions, i.e. questions that encourage and necessitate an extended form of response. Students can learn that comparatively straightforward and simple open-ended questions require particular written responses if they are to be answered effectively. Writing frames (pages 31-32) provide a good starting point for this type of work.

Next we need to consider what, from the student's viewpoint, he or she could reasonably be expected to write, given the task set. If closed questions are asked too regularly as the focus for writing tasks, students may find it both conceptually difficult, and structurally taxing, to escape from the usual 'one word answer' syndrome when they are attempting to produce a piece of extended writing. Merely stating that their answer to a particular question should be 'at least one side long' gives no help whatsoever! Instead it leaves the mechanisms, techniques and connections necessary for producing high quality extended writing unclear. Only when the steps towards the production of extended writing are mutually understood, and the means of achieving it become readily apparent and easily recalled by the students, will they succeed in creating analytical, explanatory and purposeful extended writing in geography.

The second question is more complex. The link between writing and learning in the geography classroom requires the support of research evidence to help us understand the processes at work when students 'write to learn'. The next chapter addresses the ways in which geography teachers can use the results of this research (and recommendations from it) to help students to improve their extended writing skills, and therefore their learning, in geography.

Photo: News Team International Ltd.

2: Extended writing and assessment

Geography teachers can employ a number of techniques to improve the quality of their students' written work. Chapters 3 and 4 will provide valuable practical examples of techniques that will help you enable all students – regardless of their perceived levels of achievement and ability – to produce written work that is purposeful, grammatically secure and contains good geographical knowledge and understanding. However, a research perspective is necessary to enable us to appreciate why the techniques in Chapters 3 and 4 are thought to work. Without such an understanding, these techniques are difficult to apply successfully in other contexts within geography education. However, this chapter locates the value of extended writing in relation to the opportunities it offers for the assessment of students' understanding, knowledge and skills in geography.

One of the main problems we face is that, at present, the production of high quality extended writing does not seem to be highly valued in geography at key stage 3, at GCSE and (some would argue) even within A- and AS-level examinations. Despite the demand for the completion of externally assessed coursework, the emphasis on students having to produce purposeful extended writing has steadily decreased over recent years. Geography examination papers now tend to require students to write single sentence answers or, if more marks are to be awarded for a more complex answer, space is given for a three- or four-line response. Only in the (proportionally fewer) 'higher order' questions are students expected to engage in anything that approaches extended writing in the generally accepted sense. Often students are expected to write a concise, transactional set of short sentences or even bullet points to ensure that they convey essential geographical content. The reasons for this relate to assessment. It is 'safer' (in the sense of maintaining reliability) for an examiner to mark a one-word or one-sentence answer than to assess a more complex piece of extended writing. Little, if any, opportunity is given for the type of extended writing that encourages students to formulate a structured argument, to be analytical, to reach conclusions or to display their ability to reason. Many of the classroom activities that students now undertake using geography textbooks (and other published materials) also tend to encourage this narrow approach to writing. Unfortunately, the 'double-page spread' with text, supporting photographs and diagrams, brief comprehension questions and a summary passage for the students to copy, typifies many contemporary geography resources.

So why is it important for students to engage in extended writing in geography? If all examiners need are 'one-word answers' to assess what students know, understand and can do, why ask them to produce more writing? Here are two possible reasons:

- First, extended writing compels students to support or justify what they want to say. A one-word answer to a closed question tells us little about the depth of that student's understanding; without the supporting evidence of extended writing or a debriefing session (Leat and Kinninment, 2000) we have few clues to his or her reasoning.

- Second, extended writing tells us more about how a student is thinking. It provides a key to how he or she understands and develops concepts, how gaps in that student's understanding occur, how his or her critical thinking is developing. In short, it provides us with clues as to what the next educational steps should be for the creation of that student's greater knowledge and understanding.

Extended writing can be a key to formative assessment. It can provide us with a knowledge of individual student performance on which to base effective feedback and the opportunity to feed forward to enhance future learning. Recent research suggests that the effective use of formative assessment can have a major influence on student learning, particularly when this assessment clearly links to the learning objectives that have previously been shared with them (Black and Wiliam, 1998). The standards expected for the students' work must be clear. At the end of the assessment process your students should be fully aware of their next educational steps (and how to take them), as well as being able to recognise realistic targets for their future achievement.

Figures 1(a) and (b) outline part of the formative assessment process that might be undertaken using pieces of extended writing. Both pieces of work were produced by year 9 students, to which the teacher has added a variety of annotations (the annotations would *not* be given to the students in this form). The annotations are a guide to the formative assessment that is being carried out and show how we can 'read the writing' for educational purposes. Here the teacher has considered evidence within each student's writing of:

1   reasoning and justification of points included,
2   understanding of geographical concepts, and
3   gaps within expected knowledge and understanding.

The teacher has also provided a statement of the next educational steps for each student to take below each piece of writing.

How do we get students to produce high quality extended writing? The straightforward answer might be for us to go back to getting students to write more essays, but like most simple answers this does not convey the whole truth! Good extended writing in geography does not develop of its own accord. It requires careful structuring, scaffolding and repeated practice within the classroom. As Counsel explains, in the context of history education:

'The challenge of helping students to hold on to more than a couple of propositions in their heads at once and do some "joined up thinking" is rarely addressed. Longer and more open-ended activities abound but students are

expected to leap over the abyss of structure, organisation and genre. Not surprisingly, many just fall into the abyss and never get out. Lower-attaining students, by being given over general instructions to "use the sources", or "answer the question" or "plan your answer" are simply being invited to fail. Some teachers therefore conclude that lower attainers cannot construct written analyses or explanations. But they can' (1997, p. 7).

We should not view all writing simply as an assessment 'end point': the mere culmination of what students know, understand and can do. Most writing is (or should be) a formative and educational activity. The title of the next chapter 'writing to learn' implies that writing can be used as a pedagogical tool to help students to clarify concepts, make links between ideas and engage in more advanced forms of thinking and learning. It also offers pointers on actually helping students to develop extended writing skills.

Figure 1a: Extended writing produced by a year 9 student from an 'upper band'.

Actual date of the last major eruption of Vesuvius was 1944

Not entirely consistent with the article which states that a warning 'rumble' was heard!

Some confusion over the meaning of the term 'region'. Why is the plural use here?

Who are these 'expert researchers'? What are their correct titles? How did they know the rumbles were 'false alarms'?

This section does not follow a logical line of reasoning. If the researchers 'did not trust Vesuvius', why did they tell people its rumblings were a 'false alarm'?

What does this term mean? Why is this the 'most dangerous kind of activity'?

All of them?

How many?

THE BRITISH EXPRESS.

Friday 13th October 1940
Reported by: Meera Parmar

55p

The Monster has awoke again without warning.

Mount Vesuvius the largest, active volcano in Europe has erupted yet again

Mount Vesuvius is located in the regions of Italy, between the cities, Pompeii, and Herculaneum. These ancient cities have suffered the wrath of Vesuvius before and the ruins still exist today.

The eruption started off early in the morning about 7o'clock, at the sound of a low rumble. Smoke had already started rising out of the volcano, and the people who were close enough to realise the activity became frantic and had the courage to warn others of the danger and evacuate the site.

The rumbling had carried on through the afternoon and expert researchers had declared the rumbling a false alarm, and that it was only a minor eruption and was safe, so the people of Italy who lived close to the active volcano had nothing to worry about. Although researchers did not trust Vesuvius, because Vesuvius is a so-called 'strato-volcano'. A 'strato-volcano' congers up the most dangerous kind of activity a volcano can do, and can become the most harmful.

At about 6 o'clock in the evening when the people of Italy were at ease, there was a very enormous rumble and the sound of crackling could be heard at the sound range of at least 100 km away. The people of Italy came to a panic state. This eruption was not at all expected. The lava had exploded out of the volcano bringing with it ash, smoke, and volcanic bombs, which showered miles around the area, that Vesuvius covered and further. The lava poured down the steep slope slowly, but dangerously for the millions of people at the other end.

The people who were affected mostly were the people without transport and the animals who were forgotten about. The tragedy of this loss shall always be remembered for their courage.

The plates caused the eruption according to the researchers, under the surface of the earth that are broken into several pieces. When these plates collide with each other, they cause a SEISMIC ACTIVITY.

Meteorologists who are currently surveying Vesuvius have started to draft out possible ways of; diverting the lava flow, without hurting people, control the eruption, (notifying the next possible eruptions.)

This sentence does not make sense.

Precise process of plate movement is unclear from this. Is this the only form of 'seismic activity'?

Correct term? Does she mean 'vulcanologists'?

Note: This piece of extended writing effectively conveys the story of the eruption, has a sense of audience and is reasonably structured and presented. The next educational steps for this student might be to try to:

1. Alter the writing style from descriptive to analytical. Consider how to present evidence, follow a line of argument and reasoning.

2. Be clear about geographical knowledge and understanding. Many processes are not fully explained.

3. Clarify the terms used. Does she know what all of the terms she has used actually mean (e.g. meteorologist should be vulcanologist or seismologist, seismic activity, strato volcano)?

Section titles structure the writing – but are the sections in the most appropriate order?

Structure and organisation: straight into a list of facts about 'Hurricane Mitch' with little scene setting.

Do we have any context for this comment about the destruction of books? This is odd information to lead a description of the damage done.

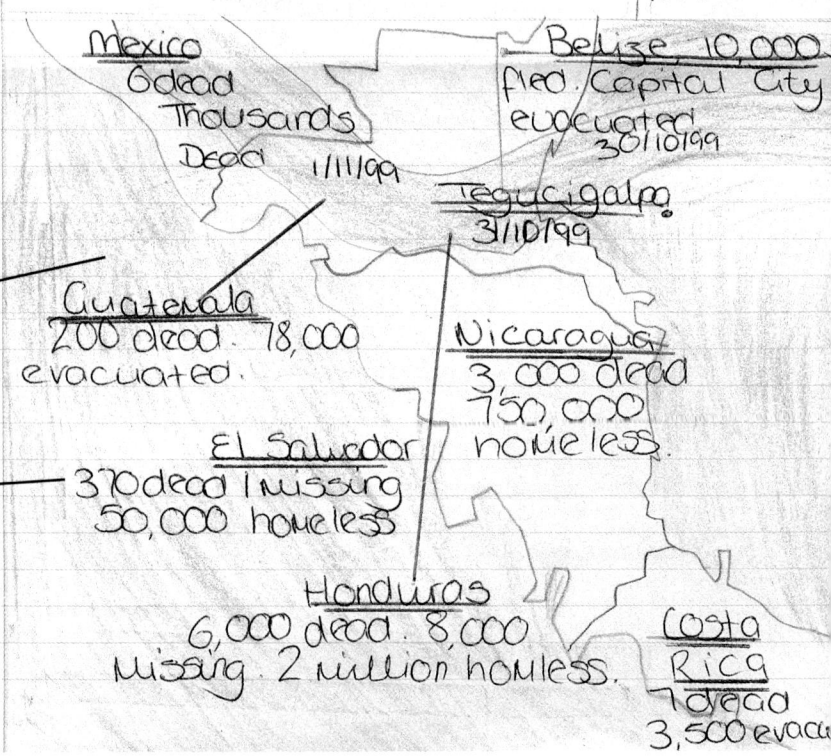

How clearly is this information conveyed? Possible use of a table or inclusion in the extended writing?

Where are the map conventions – scale, north pointer, title?

What !

The speed of 'hurricane Mitch' was well over 200 mph. The hurricane lasted 33 hours. The hurricane's power was that strong that it cost $10 billion worth of damage, nearly 4.7 million textbooks, and 4.3 million workbooks had to be ordered as the other books were all destroyed.

Where !

HURRICA

Mexico
6 dead
Thousands
Dead   1/11/99

Belize 10,000.
fled. Capital City
evacuated
30/10/99

Tegucigalpa
3/10/99

Guatemala
200 dead. 18,000
evacuated.

Nicaragua
3,000 dead
750,000
homeless.

El Salvador
370 dead 1 missing
50,000 homeless

Honduras
6,000 dead. 8,000
missing. 2 million homeless.

Costa
Rica
7 dead
3,500 evacu

Note: A good attempt at extended writing. Many facts are included, a useful attempt at describing processes and an interesting structure and use of illustration. The next educational steps for this student might be to try to:

1. Give greater context and 'scene setting' to points made.

Figure 1b: Extended writing produced by a year 9 student from a 'lower band'.

~~How?~~ Why?

Hurricane Mitch started over sea (as per usual). Cool air blew over the Caribbean Sea, and the ocean was at least 27°C, so a tropical storm started, but then the speed of the winds reached well over 75 mph, so then the tropical storm turned into one of the most devestating hurricanes ever... HURRICANE MITCH.

A good attempt at describing why a hurricane forms, but the depth of this knowledge and understanding may need to be probed.

Is this regionally or globally? Need to be more specific.

When?

MITCH. Hurricane Mitch struck at the end of October 1998. It effected countries in Central America, before hitting Mexico and Florida. It was described as the ~~most~~ destructive hurricane in the past two-hundred years.

Is this the most appropriate structure for this last section?

~~Stop?~~ How?

Yes, the hurricane effected over 2½ million people. 20,000 killed, 2 million homeless.
It destroyed lives, homes, schools, buisnesses and other things that the public used.

Vague sentence, be more specific.

2. Think about which information should be grouped together, i.e. all points about scale of destruction to appear together, and what order the points should be made in.

3. Consider what the reader may not already know. Do not assume that the reader knows what you know.

Photo: News Team International Ltd.

3: Writing to learn

To produce high quality extended writing students will have to think clearly about a variety of geographical facts, concepts and ideas and then put these together into a framework that makes sense both for themselves and for the readers. They should also think about the intended audience, i.e. those peole who will read what they write (sometimes referred to as the 'implied reader').

By making students 'move around' ideas and concepts before they write, clarifying the points for themselves and others and possibly making new links between them, we are also increasing their comprehension of geography. Students will have to consider how an extended writing task is structured – both during the planning phase and the actual writing – and this ordering activity should increase their understanding of geography before they communicate to the reader what they know, understand and can do. However, the 'abyss' that Counsel (1997) mentions above cannot be crossed without a variety of educational ladders, ropes and safety nets! You will need to provide support and guidance before students will be able to do this on their own.

Too often we assume that the writing process is straightforward and simple. Selecting relevant geographical points, analysing how they fit together, reconvening them into an extended answer, and then successfully writing that answer requires a series of logically structured and correctly sequenced steps. Most of the extended writing tasks we set do not acknowledge this and we often expect students to go straight from the first stage of information gathering to the last stage of successful extended writing in one 'leap'. Occasionally we may concentrate too heavily upon 'the geography' and too little upon the process of students' writing. Without our support students will be encouraged to 'lift' geographical information from (usually) one source and write it down.

Approaching extended writing

Students initially find that achieving high quality extended writing in geography is difficult because they have to 'carry' large amounts of information in their heads (and/or refer to a wide range of different sources) as they attempt to organise their writing. The 'steps' involved in transferring content into writing are numerous and have to be repeatedly practiced before students are able to carry out the process unaided. This activity can be complicated by the fact that:

- new information may be revealed that somehow has to be included,
- old information may need to be rejected in the light of new evidence,
- a new line of argument or analysis may need to be established, and
- some information may not easily 'fit' within the emerging answer and therefore may have to be rejected or reformulated.

At the same time students will need to consider the 'bigger picture' in a piece of extended writing. For example:

- What are the 'big points' that must be emphasised?
- What are the relatively insignificant pieces of information that still need to be included in support of the bigger picture?

Such complexity of structure, organisation, genre, audience and analysis is daunting for even the most experienced writer. For the least experienced or less able student these challenges often result in him or her following the easier path of simply copying information or including irrelevant 'padding' in an answer.

Counsel (1997, p. 13) outlines the main problems students face in creating high quality extended writing by concentrating on: memory and construction, relevance and selection, sorting, general and particular points and the language of discourse. Each of these five aspects of the writing process is dealt with separately below.

Memory and construction

Students often need to be offered opportunities to develop the skill of retaining more than one important point in their short-term memory, while making decisions about the status of such pieces of information, and finally attempting to relate these points to each other before writing. Within the context of answering a specific geographical question (or questions) achieving each of these interlinked steps can prove to be problematic for many students.

Relevance and selection

Students can often achieve a workable idea about what is and is not relevant information to be included in an answer. However, they need to develop the means of both selecting, and justifying the selection of different facts or points. Creating a set of basic criteria for the selection of information that will eventually be included in their writing is an important reasoning activity.

Sorting

The process of sorting pieces of information establishes their relevance to a particular question. Classification and sorting are often closely linked; with sorting being a pre-requisite to eventual description, analysis and evaluation. The need to establish patterns and order when sorting, to label information correctly and to use key geographical terms to help in the sorting process, achieves a variety of educational goals. Sorting helps to develop students' thinking skills and is an important stage in marshalling information before they attempt to write.

General and particular

Often students have difficulties in seeing the difference between those points that are general and those that are particular. They need to practice using geographical evidence to support a position taken, defining which are the 'small points' and which the 'larger points' in an argument, and to be clear about which specific points can be raised as generalities. Many students need also to be aware that it is usually not possible to substantiate a general conclusion with a single piece of specific evidence.

The language of discourse

Students often need to develop greater sophistication in the form of their writing, and therefore need to be exposed to a variety of texts. They need to be offered opportunities to create different types of prose, experiment with different 'starters' and 'finishers' for sentences, create connections between sections of text and confidently use simple causal connectives (e.g. 'therefore', 'and so', 'thus', 'as a result'). Indeed, students need to learn the function and use of these and more complicated connectives (e.g. 'despite', 'notwithstanding', 'although') within geography texts.

The following chapter offers practical ways of helping students to 'transform' a variety of information into a more thoughtful, extended, analytical piece of writing. The techniques are offered in the context of a series of small research projects undertaken in the West Midlands, concerned with enabling students to engage in the production of extended writing in geography.

Cartoon: Dave Howarth.

4: Achieving high quality extended writing

The research base

A series of small action research projects, concerned with enabling students to engage in the production of extended writing in geography, were conducted with year 7, 8 and 9 students. Each year group undertook a variety of 'intermediate' oral and written activities before they engaged in producing a piece of extended writing. As an article written on part of one of the research projects stated:

> *'It was felt that the students' extended writing might be enhanced by consolidating their initial geographical learning through introducing "intermediate" teaching strategies. Evidence from earlier work suggested that many students required more time, and more structured activities, to help them make the step to successful (audience-centred) writing. Thus in the lessons immediately before the extended writing task students were given "intermediate" tasks to complete. Each of these was designed to help students to develop their conceptual understanding of the geography taught, and to appreciate the kinds of information they would be required to use in their subsequent writing tasks'* (Butt, 1998, p. 207).

The techniques are not new (a substantial amount of literature is readily available on them), however, each one was used as part of the project to verify theories about how students approach writing tasks. Not all of the techniques are uncontroversial – evidence exists that in certain situations not all teachers and students have found them supportive to the process of writing. However, they bear further consideration when supporting students in their production of extended writing and (suitably adapted) can be used successfully with any age or ability range.

The following sections detail the 'intermediate' activities and an end-of-unit extended writing task undertaken by year 9 students in one of the comprehensive schools. Generally, the students worked through a unit of work on 'Ecosystems', and specifically they looked at tropical rainforests. Over the 14 weeks of the project, the students practiced sorting, analysing and restructuring geographical information, both orally and in writing. The activities were designed to help students achieve a fuller comprehension of

the geographical concepts to which they had been introduced in previous lessons, so not all ended with the completion of a piece of writing. The most important aspect was to make explicit to the students that what they were doing was structuring their thoughts, so that when an extended writing task was eventually introduced they could refer to techniques they had learned previously and undertake it more successfully. On occasions the teachers and/or researchers added a further piece of information (e.g. a newspaper article, a photograph, a passage of text and/or diagram from a textbook) to enhance the students' understanding of the geography. However, we did not want to swamp students with too much information or 'low order' writing, such as copying text, completing one word or single sentence answers, or engaging in simple comprehension. Our overall aim was to get them to handle the geographical information and mentally 'play' with it, to start to analyse and question it, to make it part of their own geographical understanding and knowledge, rather than merely to 'transact' it onto a page for the teacher to mark.

Often the teacher/researcher discussed with individuals, pairs or small groups of students how they were carrying out different tasks. In this way he or she tried to encourage 'thinking about thinking' (metacognition) so that the students were actively considering how they were completing tasks or, where they encountered problems, how they might resolve them if they encountered similar problems at a later date. (Similar techniques can also be found in Scardamalia *et al.*, 1981 and Counsel, 1997.)

The activities

Four examples of the 'intermediate' activities – ordering, card sorting, writing frames and Directed Activities Related to Texts (DARTs) – are reproduced below.

Ordering

Ordering activities help students to handle data and information confidently before attempting to engage in writing activities. They mirror some of the processes that writers go through before attempting to construct a piece of extended writing – that is selecting pieces of information, isolating key points, restructuring facts into an order that makes most narrative sense, and presenting a final piece of writing that has a clear direction and a message within it. By undertaking ordering activities students learn that their own extended writing can be successfully structured through correctly piecing together a series of geographical concepts, ideas and facts which then link together to form a meaningful whole. Ordering activities, such as that on rainforests shown in Figure 2, are only the 'first steps' in producing extended writing because they do not require students to structure a piece of their own writing, or indeed actually to *write* anything. The process is important as a signifier of what students will have to do independently, once they have gathered information from a range of sources. Issues of structure, genre, audience and form will all follow once the basic approach to ordering is achieved.

Figure 2: Example resource cards for an ordering activity. After: Bishop and Prosser, 1990.

1. Tropical rainforest is inhabited and used by Amazonian indians.	7. Lands abandoned after 15 to 20 years. Soil is now infertile, nutrients have been washed away, topsoil has been washed into rivers by rapid runoff. Scrub takes over from weeds and grasses.
Crop yields fall. Settlers abandon their cleared plots. In some areas up to 50% of the settlers move within five years of arrival. Often settlers clear new plots for cultivation.	Ranching land becomes extended. Weeds take over from grasses. Ranchers begin to clear new lands for ranching from remaining forest.
New settlers clear plots of land. Ranchers also move in to remove larger areas of forest for cattle ranching.	Roads built into the rainforests to allow access for new settlers and commercial projects (such as ranching, mining and logging). Indians often move further into the forest.
Ranchers start to take over even more land – often including plots left by settlers who have moved now that their crops have failed. Many remaining Indians have died from diseases introduced by new settlers and ranchers.	

Card sorting

Card sorting activities provide a helpful extension to the ordering tasks described above. In the example shown in Figure 3 students have to prioritise a series of uses for the rainforest in terms of how environmentally damaging each one is. In so doing, students are undertaking a structuring activity which is very similar to one that immediately precedes the production of good extended writing. By sorting out the most and least destructive uses of the rainforest, for example, the students have achieved a valuable framework for a piece of developed writing and on damage to the environment. Selecting information is the first stage of the process, followed by prioritising and organising it, then – if a writing task is actually set – students must produce a synthesis of this information.

Nash (1997), Leat (1998) and Leat and Nichols (1999) offer similar card sorting activities. These researchers specifically stress that such activities provide a route to effective group work rather than necessarily being developed into requiring individual students to produce a piece of extended writing. Activities involving sorting, labelling, ranking, and matching words and definitions on cards can all be used to promote discussion and to create a point from which the step into writing is more straight forward.

Figure 3: 'Rainforest' cards. After: Bishop and Prosser, 1990.

Shifting cultivation	Ranching
Shifting cultivation People live on the resources of the forest and only clear new plots of land every few years, leaving old plots to recover fertility. The forest is also used as a source of fuel, to supply food, clothing, medicines, etc. People rarely take out of the ecosystem more than they put back.	**Ranching** Large areas of rainforest are simply cleared to make way for large-scale cattle ranching. Trees are replaced by grassland. Nutrient supplies to the soil are reduced and grasses become weaker and fail. New lands are then often cleared to provide additional ranching areas.
Mining Roads or railway links are built in the forest in areas that have minerals that can be mined. Minerals such as copper and iron ore are often quarried by mining companies. Trees are removed to create the routeways, and to clear the area to be quarried.	**Farming** Settlers move into the rainforest, often attracted in by government schemes to develop these areas. The settlers clear land and set up farms in the hope of creating a 'new life' for themselves.
Logging Timber companies select the trees they want from the vast variety of trees in the rainforest. They may take only two or three trees per hectare, but in removing the trees they may also destroy or damage half of the other surrounding trees. Logging is a major use of the rainforest, second only to agriculture. Pulp mills are also set up to make paper.	**Plantations** Companies plant large areas with a particular species of tree which they want to harvest, rather than using those trees currently growing in the rainforest. The advantages of plantations over naturally-growing forest trees is that they contain only the trees the harvester wants, can be thinned out and used at various stages of their growth, can be sprayed to control disease, and can be planted in phases to keep yields high.

To make such activities more challenging, ask the students to sort the cards into more than one 'column'. By doing so, they should understand that the construction of a single prioritised order of information is not always the expected outcome. In the rainforest activity the basic information is given on the cards shown in Figure 3. To make the task mirror more accurately the first-hand selection of data, from which the student would then be expected to compose a piece of extended writing, other sources of information can be suggested. These information sources can range from previous whole-class discussion, to teacher-supplied notes, textbooks, photographs and the internet. With younger or less able students it may be important to restrict the range of sources to avoid the risk of the task becoming too complex and difficult.

Providing the information on cards allows students to move text around without them being in danger of writing the 'wrong thing' first or engaging in effort-sapping writing that has then to be discarded. In addition, research has shown that useful student discussion often arises from such sorting and ordering activities, which in turn provides evidence of first order thought and concept clarification (Leat, 1998). However, as Counsel warns:

> *'Without careful planning, a lot of "resource-based" or "enquiry-based" learning can lead to rampant copying, thus failing to develop the very capacities of independent, critical analysis for which it was designed ... If the goal is to help all students to use wider ranges and types of information, in increasingly independent enquiry, the teacher's attention to structure in their thinking must be correspondingly great'* (1997, p. 19).

Card sorting exercises should therefore be used by teachers as opportunities to introduce students to 'texts' from a variety of original sources, to help students to order and structure their thoughts, and to avoid the simple copying of information without comprehension of the concepts contained within it.

Writing frames

Writing frames provide students with the initial structure for a piece of extended writing. There are many forms of writing frames and they can be differentiated according to the perceived strengths and/or weaknesses of the students. They are usually designed to provide either the means of completely structuring the students' writing or to only loosely support the writing process (see, for example, Figure 4). As with card sorting activities, writing frames provide a simple way of reducing the pressure on the students to complete the variety of tasks necessary to producing good extended writing. Writing frames support and scaffold the organisational process before students start writing so that they can concentrate on making decisions about what to include and where to include it. Students are, therefore, encouraged to come to decisions about 'which facts go where' when writing in an extended fashion, rather than rushing headlong into recording lots of unrelated facts in almost any order.

Figure 4: An example writing frame on rainforests, used by students during the project.

> Complete each of the lines given below. You may write more than one line if you feel you have more information to give.
>
> When the tropical rainforest is removed rain falls directly onto the bare soil. This causes the soil on the surface to _____
>
> _____
>
> The nutrients that are in the surface soil _____
>
> _____
>
> The leaves that used to fall onto the soil from the rainforest trees no longer do so. This affects the nutrients in the soil because _____
>
> _____
>
> Runoff from the bare soil surface increases. The water that now flows into the rivers
>
> _____

A simple writing frame, linked to the rainforest work described above, is shown in Figure 4.

By getting students to order and structure the geographical information they have been presented with, often from a number of sources, the teacher can make some valuable judgements about how well the student understands the geography. Students should not see writing frames as being fixed and unchangeable – once their confidence in using writing frames grows students should be encouraged to alter frames to suit a specific purpose and even devise their own (Wray and Lewis, 1994). The writing that ensues from using a writing frame actually supports the assessment process because it gives us an important insight into the thinking processes the student employed before he or she started to write. This signifies whether the difficulties that that student faces when attempting to write in an extended fashion are related to his or her understanding of the geography, or an inability to order this understanding into a meaningful piece of writing, or both. Such assessment opportunities are formative. They help us to understand the next educational steps that students should go through to increase their geographical understanding and skills.

Photo: John Jebb.

Directed Activities Related to Texts

All of the above activities might also be described as Directed Activities Related to Texts (DARTs). Figure 5 helps to explain the range of techniques that might be applied to text either broadly for 'reconstruction' purposes (that is activities that require students to use a text that has been modified in some way by the teacher), or for 'analytical' purposes (that is an activity that gets them to use text in its original form).

Figure 5: The range of activities that can be applied to text.

'Reconstruction' activities	'Analytical' activities
Text completion	Text marking
Sequencing	Labelling
Prediction	Segmenting
Table completion	Table construction
Diagram completion	Diagram construction
	Student generated questions
	Summary

Counsel (1997) suggests that some DARTs techniques can be used to enable students to analyse both their own and other people's writing. In situations where students are asked to draft and redraft their writing, these strategies may prove particularly useful in suggesting to them where changes may be necessary. They also help the students to understand the importance of order and structure in their writing. In this way students' writing can be used as a teaching resource in a truly formative way.

The following techniques are based on suggestions made by Counsel (1997, p. 38).

Try using the instructional phrases to direct students reading of their own writing ...

1   Shade all the points you have made about x in green, and all the points about y in red. What do you notice?
2   Underline all the points where you directly refer to an information source. Look at   how you introduced each point. Compare your wording/writing with your neighbour's.
3   Cut up your writing. Make cuts between each paragraph. How easy is it for your neighbour to arrange your paragraphs correctly?
4   Find all your 'big points' and underline these in red. Are there places where you needed more 'little points' to support these?

... and on work that has been marked, write:

1   Read all of my comments in the margin. Can you think of two targets for next time   you do a similar piece of work?
2   I have underlined four things on this page. Can you work out what they have in common?
3   Show me the place in this writing where you stopped writing about the causes and went off the point.
4   With which parts of your writing do you think this geographer (quotation) would not agree?

See, also, the work of Westoby (1999) on text sequencing, construction, completion, annotation, and use of DARTs.

Extended writing tasks

After the year 9 students had undertaken the various 'intermediate' tasks on the theme of tropical rainforests described above they were given a choice of one of three tasks. They were asked either to write a structured essay on the subject or to compose a letter to the United Nations calling for further protection of the rainforest or to construct a script for the television series *This Fragile Earth*.

Concentrating upon the essay, as the most usual form of extended writing expected of students, the task set was:

> Write a structured essay to answer the following question: 'What are the main threats to the world's rainforests and would their destruction be a serious loss to the global ecosystem?'

The work produced by the students revealed the benefits of asking them to complete ordering, sorting and framing activities prior to writing an extended essay. Many students produced high quality transactional writing. Those students who had not, in their previous work, been identified as particularly 'strong' academically, appeared to benefit from the practice they had had in structuring and ordering geographical information. Their extended writing pieces indicated that they appeared to have repeatedly used the techniques they had learned. Often their writing was balanced and geographically accurate, combining a number of relevant points into a logical structure. One student, Claire, produced a series of well researched paragraphs which started with 'key sentences' to order the extended writing. Claire's 'starter' sentences were as follows:

- 'The tropical rainforests are important because ...'
- 'The problems that the rainforests face are ...'
- 'When the rainforests are destroyed ...'
- 'The ways in which we can help ...'
- 'I feel that it is important that governments take action because ...'

The work revealed that Claire had entered into a structuring activity which helped her to convey the key geographical points she wished to make. These showed Claire's knowledge and understanding of the topic, managed to include her values and attitudes, and revealed aspects of her political literacy. A second student, Donna, after a scene-setting paragraph, structured the essay under the following subheadings:

- 'Why are the rainforests under threat?'
- 'What are the consequences?'
- 'The benefits of keeping the rainforests'
- 'The natural habitat and its protection'
- 'The solution'

The extended writing of both these students, and of others within the group, showed evidence that the previous techniques of structuring, ordering, sorting and analysing information had been learned, at least to some extent. The fact that these techniques had been successfully applied to the production of extended writing, with little or no prompting from the teacher, was extremely encouraging.

Photo: John Jebb.

5: Broader considerations

This chapter considers the ways in which teachers and researchers have promoted different approaches to using resources, considered different types of discourse and encouraged students' thinking about audiences to produce extended writing in geography.

Brownsword (1998) and Bermingham et al. (1999) have helped broaden their own ideas about the production of extended writing by leading students towards deeper considerations of what constitutes 'good quality writing'. Brownsword (1998) is concerned that students' writing should reveal empathy for non-western cultures and has developed a range of language-based activities designed to promote this. Importantly she stresses the need for students to use both expressive as well as transactional forms of language and briefly focuses on the nature of audience in writing tasks. The range of activities Brownsword (1998) outlines is designed to help students to explore ethical, social and citizenship issues and therefore encourages them not to write solely from a factual, transactional perspective but to engage in using expressive, values-related and affective perspectives. Persuading students of the importance of writing about their own views and reflections, or simply encouraging them to make comparisons between their own and other people's lifestyles in their geographical writing, can be instrumental in helping them to produce high quality work. On a related subject, Bermingham et al. (1999) remind us that all written texts have been produced from a particular perspective and that students should read, analyse and aim to produce writing from various stances and for various purposes. Here the term 'text' is used in its broadest sense – from the written word to videos, pictures and sketches. According to Bermingham et al. (1999) reliance on only one text is seen to be limiting due to the partiality of perspective that results.

McPartland (2001) discusses the use of narrative discourse within the context of using moral dilemmas to help teach geography. McPartland argues that a narrative structure helps students to understand geographical concepts partly because the statements in the discourse are usually arranged in chronological order, have a person recounting the story who is often a central participant in the narrative and tie events in the past to actors in the narrative. Leat (1998) makes a similar point when he considers the importance of a narrative form within the construction of effective Mysteries. He argues that letting students create such narratives for themselves can help them to order and structure geographical events and concepts.

Other forms of discourse also exist, namely *expository discourse* (which explains concepts), *argumentative discourse* (which allows actors to state a case and defend a position), and *procedural discourse* (which instructs someone how to do something). Students will benefit from being given opportunities to use these different forms of discourse in some aspect of their production of extended writing in geography.

It is increasingly clear that good quality extended writing can be produced by students if they are offered a variety of supporting activities which will help them to improve their writing techniques and enhance their understanding of geography (such as those outlined in the previous chapters). Asking students to produce different forms of extended writing for different audiences will increases their repertoire of writing styles, it also makes them consider the most appropriate ways of marshalling geographical concepts or facts. These ideas are not new. The Writing Across the Curriculum Project in the 1970s (Martin *et al.*, 1976) and the National Writing Project of the 1980s (Richmond, 1986) outlined the need for students to be given a range of writing tasks. They also highlighted the need for students to be given opportunities to produce writing:

- of different kinds (e.g. personal, descriptive, creative, discursive, analytical),
- in different forms (e.g. poetry, plays, newspaper articles, letters, stories), and
- for different audiences (e.g. teacher, other students, friends, parents, newspapers, firms).

Both Projects also sought to encourage more 'open' writing where students could:

- write in their own words, at length,
- have opportunities for drafting and rewriting,
- have opportunities for collaborative as well as individual writing, and
- expect selective and sensitive marking by the teacher, encouraging students to become their own assessors.

The signs of improvement in extended writing

The production of good quality extended writing in geography does not occur overnight. It is apparent that students pass through a number of stages from producing poorly structured, concise, incomplete writing to creating writing that is expressive, balanced and analytical. Lewis (1989) helps to define the common characteristics of the earlier and later stages of such writing. These characteristics are shown in Figure 6.

However, care is needed when applying such definitions to students' writing because not all individuals exhibit similar characteristics and problems at the stages suggested above. Even quite accomplished writers, for example, may continue to have problems regarding, say, the use of the apostrophe or may assume that the 'implied reader' has more background knowledge of the topic/subject than their text provides.

Figure 6: Common characteristics of the earlier and later stages of writing. After: Lewis, 1989.

Earlier stage	Later stage (in some kinds of writing)
Simple sentence	Complex sentence with subordination and embedding
Connectives: 'and' 'when' 'then'	Connectives: 'so' 'because'
Active voice	Passive voice
Personal, using 'I' 'we'	Impersonal, using the third person
Narrative with sequencing by time	Scientific genre with logical sequencing
Present tense and regular past, e.g. with -ed endings	Irregular past tense, use of more auxiliaries and more complex verb form
May assume teacher knows a lot of the background	Assumes that text stands on its own and may be read by anyone
Repetition of same sentence structure and often the subject noun repeated in different sentences	Varied sentence structure and use of connective devices between sentences
Unsure about the use of capitals and full stops	Sentences accurately marked
Unsure about the use of apostrophe for possession and abbreviation	Correct use of apostrophe for possession and abbreviation
Little use of adjectives	Wide use of adjectives
Strong influence of speech on writing	Has taken on the role of writing as different from speech

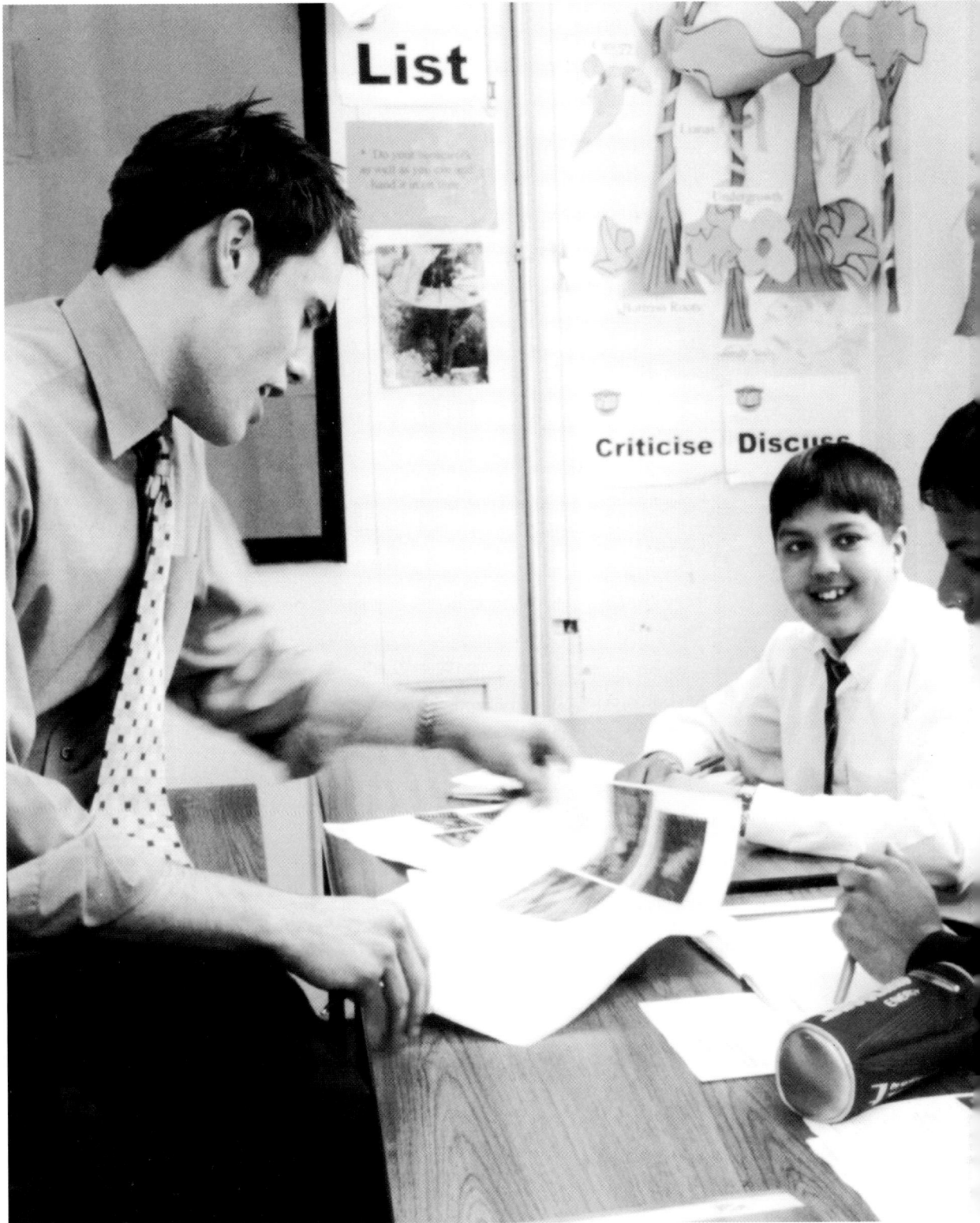

Photo: News Team International Ltd.

6: Conclusion

Although supporting students in their efforts to produce high quality extended writing in geography may (initially) be time consuming and frustrating for both student and teacher, the return on the investment can be considerable as students continue to develop and apply the skills they have learned to new writing tasks, geographical concepts and learning situations. There will always be a tension between how much support a teacher should give to a particular student who is struggling to complete a piece of extended writing, and how much he or she should be allowed to gain independence in ordering, structuring and framing the writing task. This is part of teachers' professional judgement and will vary with the particular context and content of the student's learning. As Counsel comments:

> *'Students need to be assured of a minimum success and helped to take considerable learning risks. The management of this tension is a key measure of the success of any teaching strategy'* (1997, p. 39).

This book provides some useful guidelines that can be applied to the whole process of helping students to improve their extended writing skills in geography. By repeating a variety of 'intermediate' activities, such as selecting, sorting and ordering pieces of geographical information (pages 27–35) students should eventually become more confident in their own abilities to structure pieces of extended writing. As they become familiar with these techniques, more advanced forms of organisation of information sources can be introduced using a greater breadth of materials, new geographical vocabulary and/or more challenging questions. Students can also be encouraged to analyse their own and other people's writing (pages 32–33), possibly as a result of considering the teacher's comments on assessed pieces of extended writing, to alert them to alternative ways of structuring and presenting text. Striking a balance between providing students with overly supportive or formulaic writing structures and leaving them to struggle with poorly directed or open-ended writing tasks is never easy. However, with increasing experience of 'what works', both teacher and student can achieve pleasing results.

The ability to write extended pieces that convey good geographical knowledge and understanding is important not only as an end in itself, but also as a signifier of the processes of thinking that students go through. Critical thinking, reasoning, analysing and decision-making can all be revealed through well-structured and purposeful writing.

Indeed, whenever the development of extended writing skills is considered in geography, a key question might be 'Will these activities help to enhance the students' thinking skills?'

As McLeod observes:

> 'Most school learning is language-saturated. Some people are still tempted to believe that those who have learned the language have taken on the knowledge, and that knowing something requires only that someone has heard something, or read it, and understood it. This theory implies that knowledge, and the language in which that knowledge is expressed, are identical. It also implies that language and thought differ only in that one is capable of making a noise or marks on paper, and the other not' (quoted in Slater, 1989, p. 14).

Too often as geography teachers we may be fooled by students who appear to have taken on the 'language of geography' but whose written work does not convey a depth of understanding that we might hope would be associated with that language. By helping students' to develop their extended writing in geography we may be provided with evidence both of their geographical understanding and of their emerging thought processes.

Bibliography

Andrews, R. (ed) (1989) *Narrative and Argument*. Milton Keynes: Open University Press.

Bermingham, S., Slater, F. and Yangopoulos, S. (1999) 'Multiple texts, alternative texts, multiple readings, alternative readings', *Teaching Geography*, 24, 4, pp. 160-4.

Bishop, V. and Prosser, R. (1990) *The Environment. Collins Insight Geography*. London: Collins Educational.

Black, P. and Wiliam, D. (1998) *Inside the Black Box: Raising standards through classroom assessment*. London: School of Education, Kings College.

Brownsword, R. (1998) 'Developing empathy through language', *Teaching Geography*, 23, 1, pp. 16-21.

Butt, G. (1997) 'Language and learning in geography' in Tilbury, D. and Williams, M. (eds) *Teaching and Learning Geography*. London: Routledge, pp. 154-67.

Butt, G. (1998) 'Increasing the effectiveness of "audience-centred" teaching in geography', *International Research in Geographical and Environmental Education*, 7, 3, pp. 203-18.

Carter, R. (ed) (1991) *Talking About Geography: The work of geography teachers in the National Oracy Project*. Sheffield: Geographical Association.

Counsel, C. (1997) *Analytical and Discursive Writing at Key Stage 3*. Shaftesbury: Historical Association.

Lambert, D. and Balderstone, D. (2000) *Learning to Teach Geography in the Secondary School*. London: Routledge Falmer.

Leat, D. (1998) *Thinking Through Geography*. Cambridge: Chris Kington Publications.

Leat, D. (2000) 'The importance of "big" concepts and skills in learning geography' in Fisher, C. and Binns, T. (eds) *Issues in Geography Teaching*. London: Routledge Falmer, pp. 137-51.

Leat, D. and Kinninment, D. (2000) 'Learn to debrief' in Fisher, C. and Binns, T. (eds) *Issues in Geography Teaching*. London: Routledge Falmer, pp. 152-71.

Leat, D. and Nichols, A. (1999) *Theory into Practice: Mysteries make you think*. Sheffield: Geographical Association.

Lewis, D. (1989) 'Writing in a humanities classroom' in Slater, F. (ed) *Language and Learning in the Teaching of Geography*. London: Routledge, pp. 39-58.

Marsden, W.E. (1995) *Geography 11-16: Rekindling good practice*. London: Fulton.

Martin, N., D'Arcy, P., Newton, B. and Parker, R. (1976) *Writing and Learning Across the Curriculum, 11-16*. London: Ward Lock.

McCarthy, M. and Carter, R. (1994) *Language as Discourse*. New York: Longman.

McPartland, M. (2001) *Theory into Practice: Moral dilemmas*. Sheffield: Geographical

Association.

Nash, P. (1997) 'Card sorting activities in the geography classroom', *Teaching Geography*, 22, 1, pp. 22-5.

QCA (1999a) *Improving Writing at Key Stages 3 and 4*. London: QCA.

QCA (1999b) *Technical Accuracy in Writing in GCSE English: Research findings*. London: QCA.

Richmond, J. (1986) 'What we need when we write', *About Writing* (newsletter) 2. National Writing Project. London: SCDC.

Roberts, M. (1986) 'Talking, reading and writing' in Boardman, D. (ed) *Handbook for Geography Teachers*. Sheffield: Geographical Association, pp. 68-78.

SCAA (1997a) *Use of Language: A common approach*. London: HMSO.

SCAA (1997b) *Geography and the Use of Language*. London: HMSO.

Scardamalia, M., Bereiter, C. and Fillion, B. (1981) *Writing for Results: A sourcebook of consequential composing activities*. Ontario: Ontario Institute for Studies in Education.

Sheeran, Y. and Barnes, D. (1991) *School Writing: Discovering the ground rules*. Milton Keynes: Open University Press.

Slater, F. (ed) (1989) *Language and Learning in the Teaching of Geography*. London: Routledge.

Westoby, G. (1999) 'Writing for reading and learning in geography', *Teaching Geography*, 24, 4, pp. 165-8.

Williams, M. (ed) (1981) *Language, Teaching and Learning Geography*. London: Ward Lock.

Wray, D. and Lewis, M. (1994) *Developing Children's Non Fiction Writing - Working with writing frames*. Leamington Spa: Scholastic.

Wray, D. and Lewis, M. (1997) *Extending Literacy: Children reading and writing non-fiction*. London: Routledge.